GROWING C[illegible]
ENERGY DEMAND

Is the World in Denial?

A Report of the Energy & National Security Program
Center for Strategic and International Studies

Authors
Malcolm Shealy
James P. Dorian

October 2007

About CSIS

In an era of ever-changing global opportunities and challenges, the Center for Strategic and International Studies (CSIS) provides strategic insights and practical policy solutions to decisionmakers. CSIS conducts research and analysis and develops policy initiatives that look into the future and anticipate change.

Founded by David M. Abshire and Admiral Arleigh Burke at the height of the Cold War, CSIS was dedicated to the simple but urgent goal of finding ways for America to survive as a nation and prosper as a people. Since 1962, CSIS has grown to become one of the world's preeminent public policy institutions.

Today, CSIS is a bipartisan, nonprofit organization headquartered in Washington, D.C. More than 220 full-time staff and a large network of affiliated scholars focus their expertise on defense and security; on the world's regions and the unique challenges inherent to them; and on the issues that know no boundary in an increasingly connected world.

Former U.S. senator Sam Nunn became chairman of the CSIS Board of Trustees in 1999, and John J. Hamre has led CSIS as its president and chief executive officer since 2000.

Library of Congress Catalog-in-Publication Data
CIP information available on request
ISBN 978-0-89206-523-0

The CSIS Press
Center for Strategic and International Studies
1800 K Street, N.W., Washington, D.C. 20006
Tel: (202) 775-3119
Fax: (202) 775-3199
Web: www.csis.org

Contents

Growing Chinese Energy Demand

Is the World in Denial?

Malcolm Shealy and James P. Dorian

Introduction and Background

Only a few years ago Chinese government leaders were optimistic that their country could quadruple its GDP between 2000 and 2020 while only doubling energy consumption. For any other developing country this would be considered an unrealistic goal, as energy consumption during development tends to grow as fast as or even faster than GDP. Yet, China had quadrupled GDP while only doubling energy use from 1980 through 2000, so government officials reasoned they could do it again.

However, China is already off track in meeting its 2000–2020 energy consumption goals, and the country is now at a point where it will be nearly impossible to prevent energy consumption from more than doubling. Energy use, instead of growing half as fast as GDP, has grown faster than GDP on average since 2000, as shown in figure 1. Notwithstanding, China's government, as well as major energy statistics agencies including the International Energy Agency (IEA) and U.S. Department of Energy (DOE), have been so captivated by the original, optimistic story line that their forecasts have not caught up to the reality of what is happening in China.

In this report we first examine the period from 1980 through 2000 and explain why it was unusual for the Chinese energy industry. Second, we look at the changes in China since 2000, including in the critical electric power generating sector. Using a simple thought experiment, we then illustrate that even with very conservative assumptions about Chinese GDP growth and income elasticity of electricity demand out to 2025, the country will experience much higher coal demand and emit much greater volumes of carbon emissions than forecast by IEA

This report is based on a presentation at the Center for Strategic and International Studies (CSIS), Washington, D.C., May 11, 2007. All opinions, analyses, and statements are solely those of the authors and do not reflect the official position of any U.S. or international organization or government agency.

and DOE. This leads to some final thoughts on the implications of future Chinese energy demand on global energy markets.

Figure 1. Actual Energy Use in China Is Well above Target since 2000

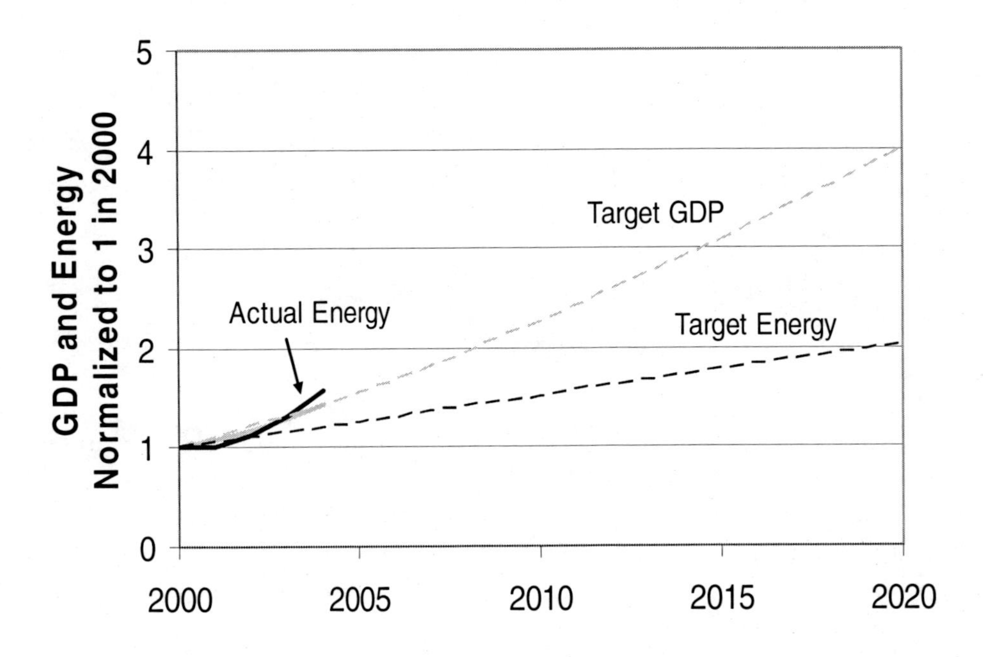

Why the 1980–2000 Period in China Was Unusual

During the 20-year period from 1980 through 2000, China quadrupled GDP while only doubling energy use. This is quite unusual. Developing country energy use typically grows faster than GDP as heavy industry develops and as consumers transition from nonmarket fuels such as firewood to market-based fuels such as kerosene and electricity.

Even more unusual for China was the weak growth in electricity demand. Electricity is the most versatile, high-quality source of energy. In developing countries electricity consumption is expected to grow faster than GDP, which implies income elasticity over 1.0. Yet, from 1980 through 2000 in China, electricity consumption grew only about 80 percent as fast as GDP, yielding an income elasticity of around 0.8. Simultaneously, reported growth in coal consumption slowed significantly in China during the late 1990s, providing Chinese government leaders with optimism that their country was somehow different and that China's economy could grow faster than energy use on a continuing basis and even transition away from coal.

Various explanations have been offered for why Chinese energy demand grew so much more slowly than for other developing countries. One prominent explanation is that the Chinese were transitioning from inefficient state enterprises to more modern and efficient means of production. The consequent efficiency improvements meant that the economy could grow faster than energy use. Another explanation is that faulty Chinese GDP or energy statistics led to skewed income elasticities. Regardless of the actual reasons, the picture has recently changed.

Relationship between Energy and Economic Growth Changed Abruptly after 2000

The Chinese economy is vastly different today than it was two decades ago, and energy linkages throughout the economy are greater than before, more costly and complex, more reliant on long-distance transport within China, and more dependent on foreign supply chains of oil, gas, and increasingly, coal. Oil use for vehicle transportation in China has grown several-fold since the 1980s, and the oil consumed is not immediately substitutable. Virtually every energy form in China during the 1980s was subsidized by the Chinese government, and energy prices were artificially low. Since the 1980s, hundreds of thousands of kilometers of new pipelines, railroads, and highways have been built, which are necessary to move increasingly larger volumes of oil, gas, and coal throughout the country.

Beginning in 2000, the relationship between energy and GDP began to change. First, the growth rate in electricity consumption rose above the growth rate in GDP for the first time in years, and continued to grow faster than GDP, yielding an income elasticity averaging 1.3 over the period 2000 to 2006. This sudden change in the nature of electricity demand may have occurred for two reasons: (1) the inefficient state enterprises have now become relatively more efficient, and as such, achieving additional energy savings from industry today is more challenging and costly; and (2) the Chinese economy has grown enormously, resulting in a smaller share of the pie for state enterprises. Today, Chinese electricity demand is behaving like that of a normal developing country and shows no sign of reverting back to the earlier pattern. Indeed, Chinese goals for rapid urbanization of upwards of 400 million people by 2030 will likely push electricity demand up even faster.

A second major change is the rapid increase in coal use starting in 2002. Part of this increase represents catching up after the underreporting in the late 1990s when the Chinese government had attempted to shut down thousands of small inefficient coal mines in order to meet World Trade Organization ascension requirements. Another reason is that coal use is increasingly driven by electricity demand. For example, in 1990, about 25 percent of coal production went to electric utilities; by 2004, that figure had grown to 50 percent. Clearly, robust growth in electricity consumption since 2000 has contributed to robust growth in coal demand—as almost 80 percent of Chinese power generation is coal fired. The increase in coal use, along with a substantial increase in oil demand, has caused Chinese energy demand to rise at a rate of about 11.4 percent per year from 2000 to 2004, while GDP growth has averaged 9.4 percent per year.

What Does China's Energy Future Look Like through 2025?

We can examine what the future may hold for China's energy industry using a simple thought experiment, and we can then compare our findings to the most recent IEA and DOE forecasts for perspective. Given that seven years have passed since 2000, we have chosen to forecast over the roughly 20-year time period through 2025. Complete historical data is available through 2004, while some data is available through 2006. Since electricity increasingly drives Chinese energy demand, we will begin with a discussion of electricity.

Electricity

To be conservative, we assume that Chinese electricity consumption grows only 1.1 times as fast as GDP through 2025, down from its current ratio of 1.3. Figure 2 shows how Chinese electricity demand would grow for two different average GDP growth rates—6.0 percent and 7.2 percent. The 6 percent GDP growth rate is close to the IEA and DOE growth assumptions. The 7.2 percent growth rate was chosen because it gives a quadrupling of GDP over 20 years and so corresponds to the targeted Chinese growth rate. For comparison purposes, figure 2 also shows recent IEA and DOE forecasts of Chinese electricity demand.

Figure 2. Thought Experiments Yield Much Higher Forecasts of Chinese Electricity Demand in 2025 than IEA or DOE

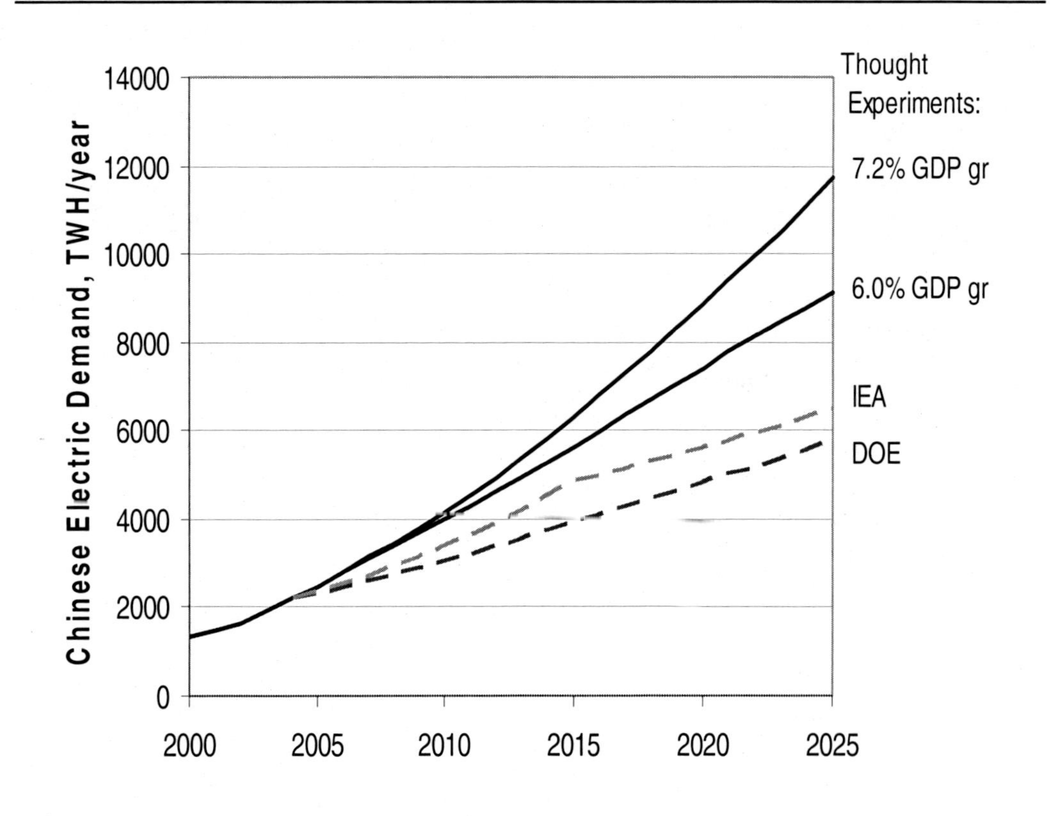

Why are the IEA and DOE forecasts significantly lower than the lowest-growth thought experiment? IEA and DOE assume low GDP growth rates, but they also assume income elasticities that are much lower than those we have witnessed recently in China. The Chinese have similarly underforecast their electricity consumption growth by using low income elasticities. As the IEA and DOE forecasters have been confronted by rapidly rising near-term electricity consumption in recent years, they have repeatedly shifted their forecasts upward (see figure 3). These forecasting problems will continue until IEA and DOE update the relationship between growth in electricity consumption and growth in GDP.

Figure 3. Like Most Chinese Electricity Demand Forecasts, IEA and DOE Forecasts Are Consistently Revised Upward

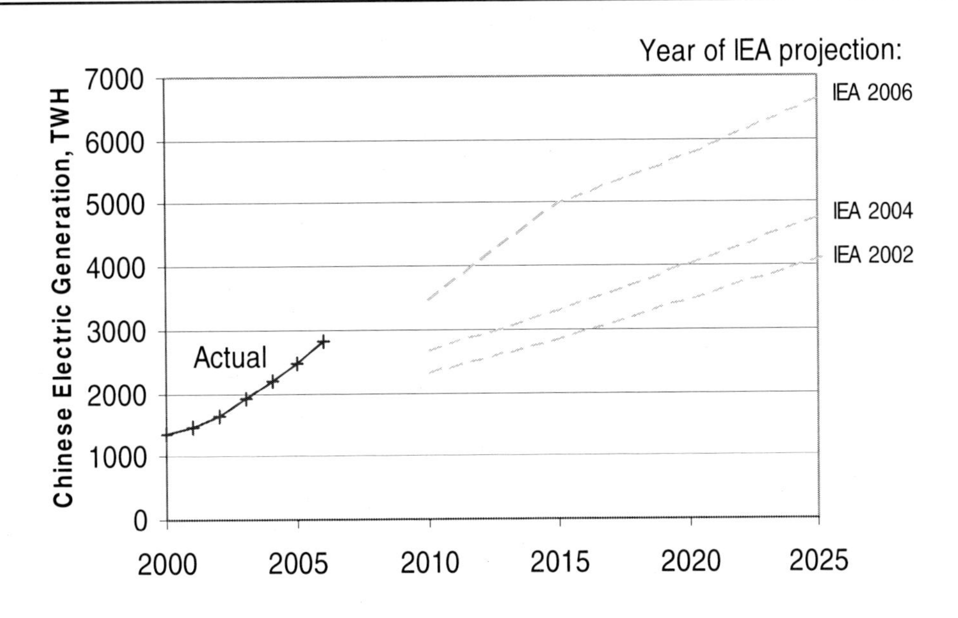

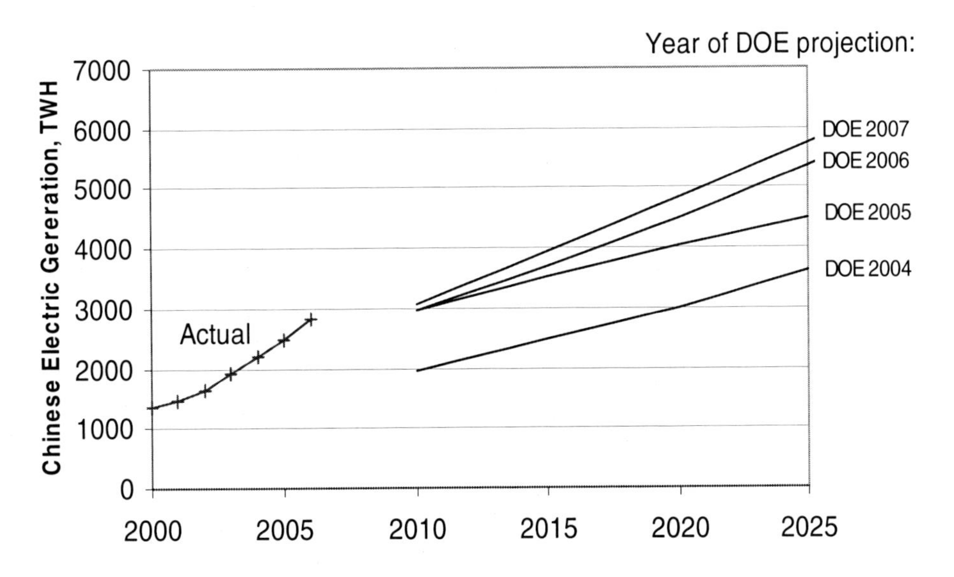

Coal

Electricity demand increasingly drives coal demand in China. The Chinese electricity sector accounted for a full 50 percent of Chinese coal demand in 2004, and its share has been steadily rising. As IEA, DOE, and the Chinese government underforecast Chinese electricity demand, they subsequently underestimate Chinese coal demand.

Figure 4 continues our thought experiment by demonstrating how Chinese coal consumption would grow for three different average GDP growth rates to 2025—6.0 percent, 7.2 percent, and 10 percent per annum. This computation is conservative in two important respects. First, it assumes more rapid growth in non-coal electricity generation than posited by either IEA or DOE (e.g., a tripling of hydro generation by 2025 and a 10-fold increase in nuclear generation). Second, it assumes that non-electric coal consumption grows only one-fifth as fast as GDP (meaning that a 10 percent growth in GDP causes only a 2 percent growth in non-electric coal consumption).

Figure 4. Thought Experiments Yield Much Higher Forecasts of Chinese Coal Demand in 2025 than IEA or DOE

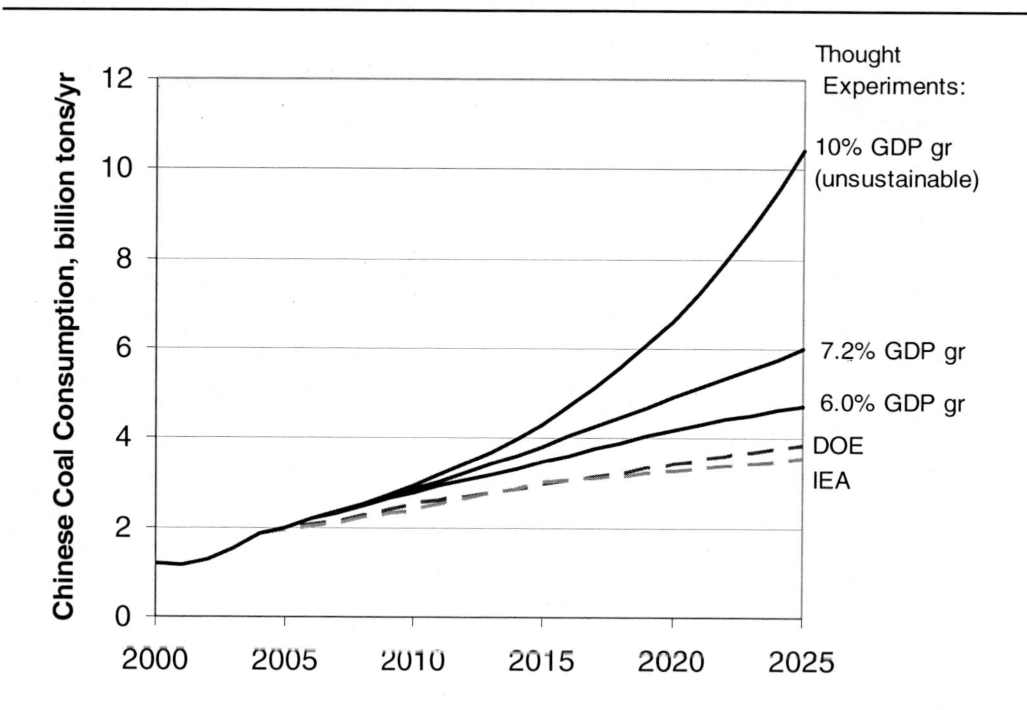

Despite conservative assumptions, with a 7.2 percent GDP growth rate in China to 2025, the country would use over 6 billion tons of coal in 2025, or three times the current amount of coal produced and used. The main culprit for this is the rapid growth in electricity consumption. To our knowledge, there are no forecasts today suggesting a tripling of coal use in China within two decades, or what the ramifications of such growth could mean to the world's efforts to combat global warming.

Figure 5. China's Coal Deposits and Major Rail Infrastructure

China: Coal Deposits and Major Infrastructure

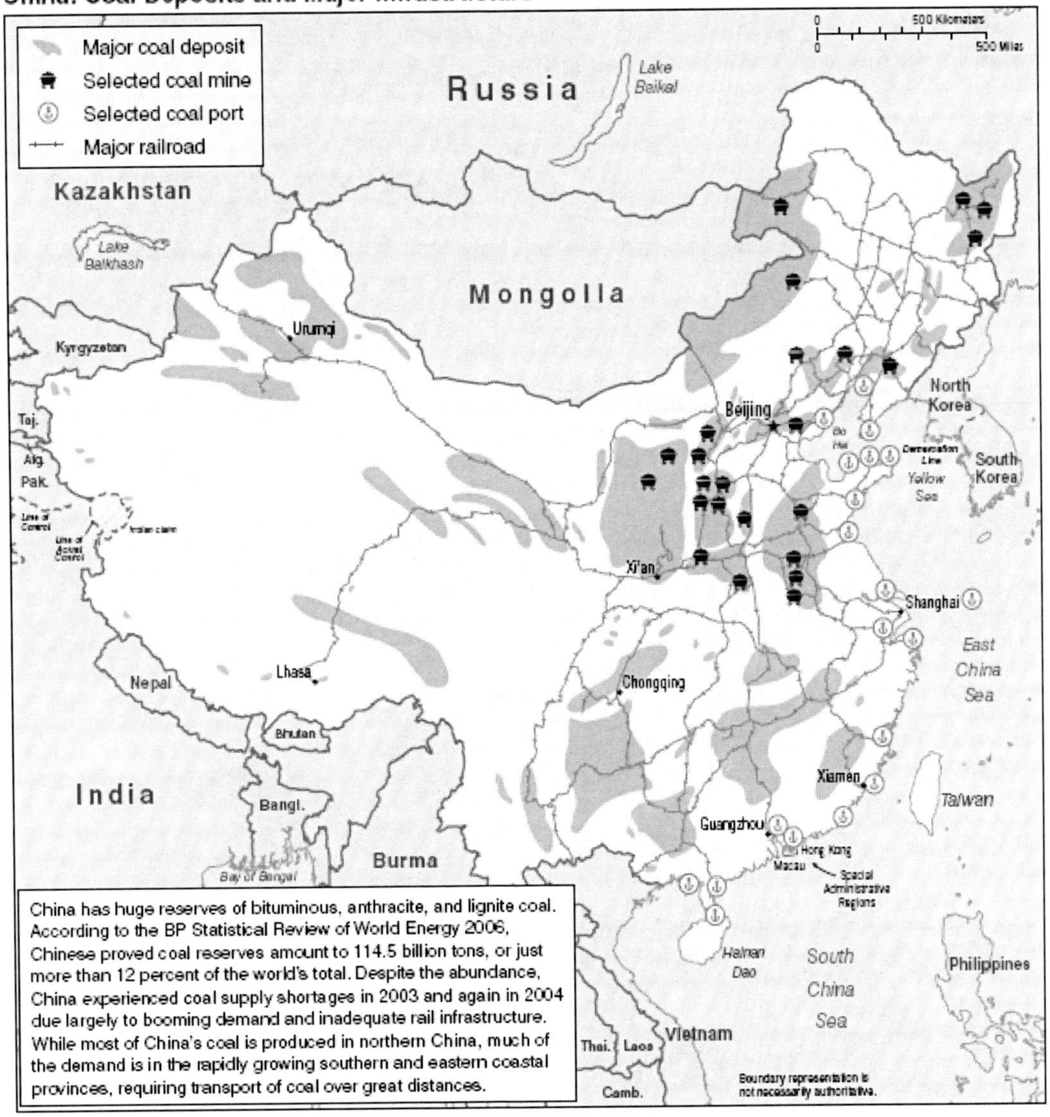
Major coal deposit
Selected coal mine
Selected coal port
Major railroad
Russia
Lake Baikal
Kazakhstan
Lake Balkhash
Mongolia
Urumqi
Kyrgyzstan
Taj.
Afg.
Pak.
Nepal
Lhasa
Bhutan
Bangl.
India
Burma
Bay of Bengal
Beijing
Xi'an
Chongqing
Shanghai
North Korea
South Korea
Yellow Sea
East China Sea
Xiamen
Taiwan
Guangzhou
Hong Kong
Macau
Special Administrative Regions
Hainan Dao
South China Sea
Philippines
Vietnam
Thai.
Laos
Camb.
China has huge reserves of bituminous, anthracite, and lignite coal. According to the BP Statistical Review of World Energy 2006, Chinese proved coal reserves amount to 114.5 billion tons, or just more than 12 percent of the world's total. Despite the abundance, China experienced coal supply shortages in 2003 and again in 2004 due largely to booming demand and inadequate rail infrastructure. While most of China's coal is produced in northern China, much of the demand is in the rapidly growing southern and eastern coastal provinces, requiring transport of coal over great distances.
Boundary representation is not necessarily authoritative.
500 Kilometers
500 Miles

Such rapid growth in coal use through 2025 would present major challenges to the Chinese economy, environment, and transportation system, and it would raise some doubt as to whether China could even meet a 7.2 percent average annual growth rate to 2025. Indeed, China's railroad cars and tracks are already overburdened, and a competition for limited rail car use has developed among coal, iron ore, steel, grains, and many other commodities, including oil (see figure 5). The increase in coal transport by rail is outpacing rail track expansion in China, displacing other freight, and accelerating a shift to less-efficient transport by road. Looking ahead, it is difficult to envision how another 4 billion tons of coal could be used in China, even with the plans for dramatic railroad expansion now called for in China's 11th Five Year Plan. Since China has become a net coal importer this year, increasing volumes of coal would have to be imported to meet a tripling of use—and it is questionable where such potentially huge volumes of coal would come from.

Oil

Oil consumption in China has risen only about 70 percent as fast as GDP, yielding an income elasticity of 0.7. However, that picture may change soon. First, the Chinese are purchasing vehicles at a very rapid rate, and the notion from a few years back that the Chinese are somehow different and will be content to ride bicycles has been disproved (see figure 6). Second, bottlenecks in rail can divert goods to long-haul trucking, which uses far more oil per ton-mile. Finally, the Chinese industrial base is a major consumer of petrochemicals and is growing rapidly.

Figure 6. Chinese Bicycle Congestion Has Given Way to Auto Congestion

Carbon Emissions

Coal supplies about 70 percent of Chinese energy needs, and this percentage is likely to remain high for decades, as coal is less expensive than many of the alternative fuels available to China. Coal, therefore, will remain the dominant contributor to Chinese carbon emissions looking ahead. Figure 7 shows the impact of growing coal consumption in China on carbon emissions, as well as growing oil and natural gas use. Once again, our thought experiments yield much higher numbers than either the IEA or DOE reference cases, even with conservative assumptions. Importantly, the Chinese could possibly double U.S. carbon emissions by 2025—China overtook the United States in 2006 according to Dutch scientists. Such expected growth in carbon emissions in China implies that worldwide efforts toward reducing global warming and greenhouse gas emissions will likely be overwhelmed by China if the country's current path of energy and economic development continues.

Figure 7. IEA and DOE Forecasts Greatly Understate the Potential Growth in Chinese Carbon Emissions

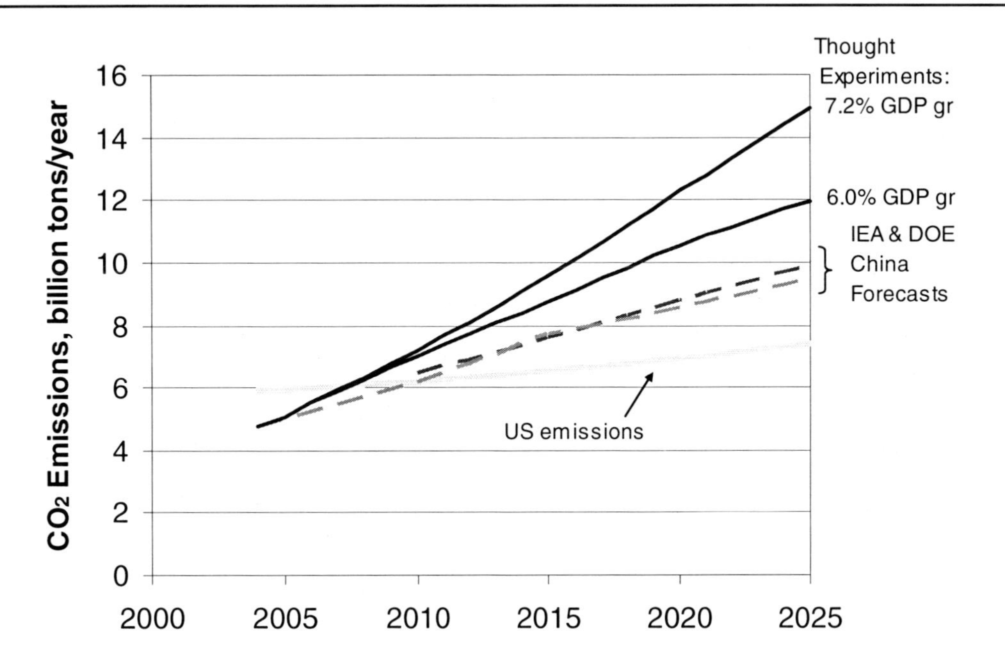

Only Way Out: Slower Economic Growth

Various alternatives have been proposed for reducing Chinese dependence on fossil fuels and slowing the growth of coal use and Chinese carbon emissions. Such alternatives have included boosting nuclear and wind power, increasing use of biofuels, rapidly developing a natural gas and liquefied natural gas (LNG) domestic industry, and increasing energy efficiency in industry and power generation through price and policy reform. Each option has limitations and challenges. However, to keep with a conservative scenario, our thought experiment assumes relatively rapid penetration of all of these alternatives—even more rapid than assumed by IEA and DOE. Nonetheless, even under optimistic assumptions regarding alternative energy forms in China, the share of electricity generated by coal through 2025 increases in the 7.2 percent GDP growth case, and remains about the same in the 6.0 percent GDP growth case. The reason is that the income elasticity of 1.1 drives electricity consumption upward at a rapid rate, one that cannot be matched by hydro, nuclear, and renewables as a group.

China has ample opportunity to increase the energy efficiency of its economy. But without dramatic and even unprecedented price and policy reform, there will likely be limitations on how much efficiency gains could slow the growth in overall energy consumption. China's reliance on low-efficiency and heavily polluting coal power plants, for example, will remain the norm without a dismantling of the huge power monopoly—which is improbable. While efficiency gains are possible throughout Chinese industry, including within the cement, iron ore and steel, and chemical sectors, high costs and lack of financial incentive have prevented significant gains to date. More importantly, China's continuing drive to achieve robust economic growth and maintain jobs— particularly at the local and provincial levels—will keep the focus of industry on output and not reduction in energy use.

With alternative fuels and energy efficiency gains likely offering only marginal help in reducing Chinese coal reliance to 2025, slower economic growth is probably the only viable solution to the current skyrocketing growth in Chinese energy demand. If the Chinese government does not reign in economic growth in a controlled manner, then various bottlenecks in the energy supply system or environmental problems will probably force them to slow their growth in a less-controlled manner.

Conclusions and Implications

As shown by our thought experiments, the IEA and DOE reference case scenarios do not represent a continuation of current trends, but instead assume an unexplained, radical reduction in Chinese energy intensity. This is evidenced by their repeated upward revisions of projections for electricity, coal, and energy in general, as well as by an unreasonably low income elasticity of electricity consumption. The actual trajectory for Chinese energy consumption is proving to be much higher than their forecasts, and this trend will continue until IEA and DOE come to grips with the high income elasticity of electricity demand.

A higher track for Chinese coal consumption has stark implications both for sustained Chinese economic growth and for the world's environment. The authors believe the world has not yet come to grips with the possibility of coal use in China increasing three-fold over today's levels, even under a conservative scenario. As such, we believe further studies and analyses should be completed to address this issue and answer fundamental questions raised by our analysis, including: (1) will infrastructure bottlenecks in the all-important Chinese coal sector lead to a slowdown in economic growth sometime next decade; and (2) how will the current worldwide escalating interest in global warming affect geopolitical relations with China, given its heavy and locked-in dependency on coal.

About the Authors

Malcolm Shealy is an energy analyst with the consulting firm PCI in Arlington, Virginia. He specializes in energy demand modeling of developing countries and also follows transportation technologies and alternative fuels. He can be contacted at malshealy@embarqmail.com.

James P. Dorian is an international energy economist based in Washington, D.C., and a former research fellow at the East-West Center in Honolulu, Hawaii. He specializes in Eurasian oil, gas, and coal issues, advanced energy technologies and alternative fuels, and the geopolitics of energy. He can be contacted at jamesdorian@yahoo.com.

About the CSIS Energy & National Security Program

The core mission of the CSIS Energy and National Security Program is to research emerging energy issues and deliver insightful analysis that educates and influences public policy. The program team is composed of individuals with a wide range of professional experience in the private and public sectors, as well as firsthand regional expertise. These experts highlight key elements, opportunities, and challenges in the changing geopolitical energy map of the twenty-first century and frame the debate over these issues in terms of relevant environmental, economic, and political imperatives. With an emphasis on strategic forward-looking thinking, the Energy and National Security Program offers realistic recommendations that help policymakers appropriately address these issues. Visit our Web site at http://www.csis.org/energy/.